THAT'S THE TICKET

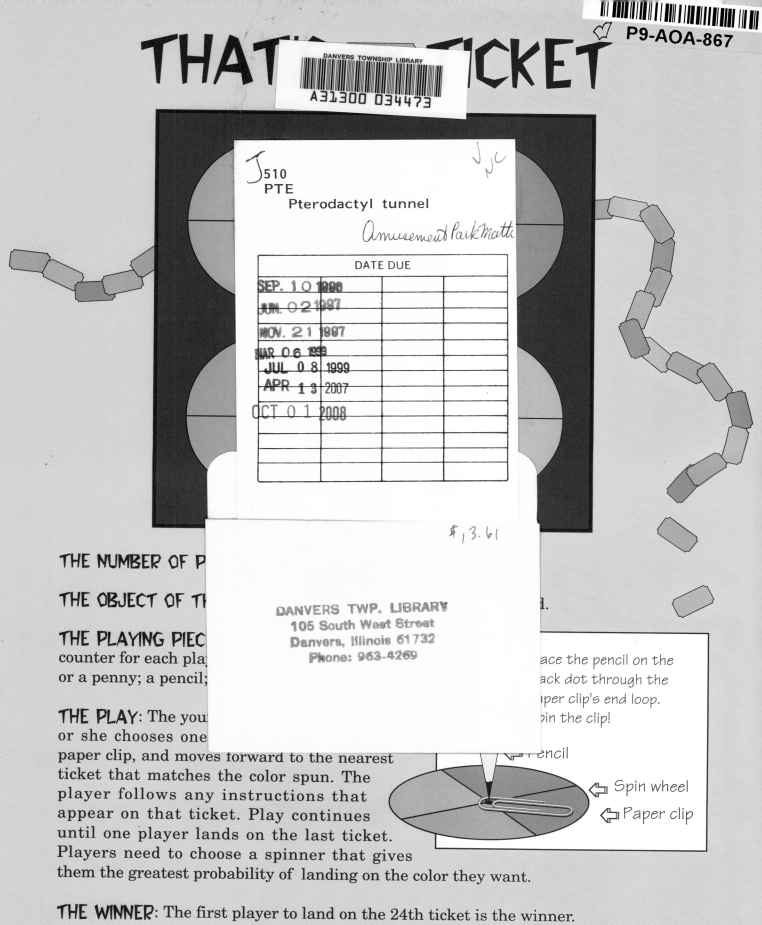

THE NUMBER OF P...

THE OBJECT OF TH... ...d.

THE PLAYING PIEC...
counter for each pla... ...ace the pencil on the
or a penny; a pencil; ... ...ack dot through the
...per clip's end loop.
...pin the clip!

THE PLAY: The you... ...Pencil
or she chooses one...
paper clip, and moves forward to the nearest
ticket that matches the color spun. The
player follows any instructions that
appear on that ticket. Play continues
until one player lands on the last ticket.
Players need to choose a spinner that gives
them the greatest probability of landing on the color they want.

⇦ Spin wheel
⇦ Paper clip

THE WINNER: The first player to land on the 24th ticket is the winner.

MATH CONCEPT: Probability.

PTERODACTYL TUNNEL

AMUSEMENT PARK MATH

TIME
LIFE *for*
Children ®

ALEXANDRIA, VIRGINIA

ALL ABOUT
I LOVE MATH

Quick! Turn to page 12 and help me find the winner of my raffle.

RAFFLE TICKETS

GRAND PRIZE
A Trip for 2
ANYWHERE IN THE WORLD!
Tickets $1.00

The *I Love Math* series shows children that math is all around them in everything they do. It can be found at the grocery store, at a soccer game, in the kitchen, at the zoo, even in their own bodies. As you collect this series, each book will fill in another piece of a child's world, showing how math is a natural part of everyday activities.

What Is Math?

Math is much more than manipulating numbers; the goal of math education today is to help children become problem solvers. This means teaching kids to observe the world around them by looking for patterns and relationships, estimating, measuring, comparing, and using reasoning skills. From an early age, children do this naturally. They divide up cookies to share with friends, recognize shapes in pizza, measure how tall they have grown, or match colors and patterns as they dress themselves.

Young children love math. But when math only takes the form of abstract formulas on worksheets, children begin to dislike it. The *I Love Math* series is designed to keep math natural and appealing.

Figure out how many more people can ride The Sugar Bowl on page 11.

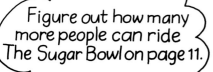

SUGAR

How Do Children Learn Math?

Research has shown that children learn best by doing. Therefore, *I Love Math* is a hands-on, interactive learning experience. The math concepts are woven into stories in which entertaining characters invite children to help them solve math challenges. Activities reinforce the concepts, and special notes offer ways you can have more fun with this program.

We have worked closely with math educators to include in these books a full range of math skills. As the series progresses, repetition of these skills in different formats will help children master the basics of mathematical thinking.

What Will You Find in *Amusement Park Math*?

Watch out for low-swooping pterodactyls as you answer math riddles and create your own thrilling boat ride through dinosaur territory. In *Amusement Park Math* you'll also learn strategies for figuring out how long you have to wait in line for a popular ride; sharpen your mental math skills as you join a couple of kids at the controls of some special video games; win a free ride once you figure out the pattern on a carousel ride; and find some unusual ticket holders at a theater.

We hope you will have fun playing the games and going on the rides in *Amusement Park Math* and will say:

I LOVE MATH!

The Editors
Time-Life for Children

How long do we have to wait to go on the Terror Tour? Turn to page 30 and find out.

Table of Contents

Thunder Park

FIND:

- a dozen-plus-2 people wearing glasses
- a half-dozen balloons on strings
- the same number of people eating cotton candy as sides on a square
- the only dog in a backpack

TRAIN

MATH FOCUS: NUMBER AND LOGICAL THINKING. By finding different quantities of objects in a picture, children use their knowledge of standardized amounts of things. Children also use deductive reasoning to solve problems.

6

Have children tell how many sides are on a square, how many objects are in a dozen, how many objects are in a half-dozen, how many fingers are on a hand, and how many days are in a week.

ALSO FIND:

• the same number of clowns as days in a week
• the same number of hats as days in 2 weeks
• the same number of flags as fingers on a hand

MORE FUN. Use logical thinking to describe something in the picture on pages 8 and 9 and challenge others to use the process of elimination to find it.

Find a woman with red shoes carrying a polka-dot bear.

Find a boy wearing a purple scarf and a green hat.

Find a girl holding 4 red balloons.

Find a man in a flowered shirt wearing sunglasses and eating cotton candy.

9

All Aboard!

Minute MYSTERY

The Viking has 2 boats.
Each boat has 10 seats.
How many people can
ride on The Viking at
once?

THE VIKING

There are 5 seats
on The Log Jam.
3 people can sit on each seat.
How many people can ride
on The Log Jam at once?

THE LOG JAM

MATH FOCUS: ADDITION AND MULTIPLICATION.
By counting groups of people with the same
number in each group, children begin to use
time-saving strategies, such as counting by 5's
and 10's, that are linked to multiplication

readiness. Children can use buttons or other
counters to make groups of 2, 3, 5, and 10 to
represent the groups of 2, 3, 5, and 10 people on
the different rides.

There are 6 cars on The Hurricane.
Each car has 2 seats.
How many people can ride on
The Hurricane at once?

There are 4 bowls on The Sugar Bowl.
Each bowl holds 5 people.
How many people can ride
on The Sugar Bowl
at once?

Which 2 rides can hold the same amount of people?

Which ride can hold 3 more people than The Hurricane?

MORE FUN. Put the same number of pennies in each compartment of an egg carton, then challenge others to figure out how many pennies are in the whole carton.

The Torn Ticket Mystery

"If I win this raffle I think I'll ask my Aunt Kitty to go with me to the Canary Islands," thought Professor Guesser as she wrote her name, address and telephone number on the raffle ticket. She tore it along the dotted line and gave that part to her friend Mr. Barns. Then Professor Guesser went off to enjoy the rest of the state fair, an event she looked forward to every year.

A few seconds later a woman rushed up to the raffle booth.

"One raffle ticket," she huffed and handed Mr. Barns a dollar.

"One ticket, it is," said Mr. Barns. The woman grabbed the ticket and rushed off. "Wait a minute. Come back!" called Mr. Barns.

"I can't. I'm in a big hurry!" said the woman. She gave the part to be filled out to Mr. Barns. "Be a dear, and fill it out for me!"

The woman hurried off. Mr. Barns tried to get a good look at her through the ticket window, but he couldn't see her face. He scribbled "red dress" on the ticket and tossed it in with all the others.

What else could Mr. Barns have written on the ticket?

MATH FOCUS: LOGICAL THINKING. Children consider more than one attribute of a person's clothing in order to solve a problem logically.

Tell children to think about size, shape, and color before deciding on the answer.

12

Later that day, Professor Guesser went back to the raffle booth to see if she had won. She tapped on Mr. Barns' window.

"Oh, Professor Guesser, am I glad to see you!" said Mr. Barns. "This is awful, just awful! The winning ticket was drawn, but we don't know the winner's name or address. There is no way to reach her, because she didn't fill out her ticket. See?" Mr. Barns handed the winning ticket to Professor Guesser.

"Then how do you know it was a she and not a he or ME?" Professor Guesser asked hopefully.

Mr. Barns replied, "The woman who bought this ticket was in too much of a hurry to fill it out. I wrote "red dress" on the ticket so I could remember her. Can you help me find her?"

270
Name.............. *red*
Address.......... *dress*
Phone Number............

the Winning Ticket

MORE FUN. Play "I Spy" by using different attributes to describe an object and then challenge others to identify it. A possible statement for a soccer ball might be, "I spy with my little eye something that is big, shaped like a sphere, and has black pentagons on it."

"There must be lots of women at the fair wearing red dresses. Can you remember anything else about her?" Professor Guesser asked.

She took a notebook from her purse and began to sketch as Mr. Barns described the Grand Prize winner.

"Yes, she was wearing a hat and she was carrying a purse," said Mr. Barns.

"Like this?" asked the professor holding up her picture.

Does Professor Guesser's picture look like the woman who bought the raffle ticket?

What needs to be changed?

14

"No, that's not right," said Mr. Barns. "The hat was orange with a wide brim, and the purse was much bigger and had polka dots on it. Oh, she also had a feather in her hat."

Professor Guesser flipped over the page and drew another picture.

"How does this look?" she asked.

Is Professor Guesser's picture correct?

What needs to be changed?

"That's closer," said Mr. Barns, "but the polka dots were purple and much bigger, and the feather in her hat was big and blue. Oh, and one more thing. She had on gloves and shoes to match her hat."

Professor Guesser drew another picture and showed it to Mr. Barns.

"That's her!" exclaimed Mr. Barns. "That's the winner!"

Is this what the woman looked like?

15

Professor Guesser looked at her watch. It was almost 2 o'clock.

"It's time for the big horse show, a very popular event," she said. "Almost everyone will be there, so it's a good place to look for our winner."

Mr. Barns and the professor stood in front of the bleachers. They could see most of the audience from there.

16

"I see a lot of women who might be the one we're looking for," said Professor Guesser.

"But which one is the winner?" asked Mr. Barns.

"All we have to do is compare each woman to the picture I drew. The one who looks most like it is the winner," said Professor Guesser and she ran into the audience with Mr. Barns right behind her.

"There she is," said the professor excitedly and she pointed to one of the women.

Which woman is the winner?

Mr. Barns checked the woman's ticket. The number was the same as the winning stub.

"Congratulations," said Mr. Barns. "Two tickets to anywhere in the world are yours!"

"Wherever you go, please send us a postcard," said Professor Guesser. "And be sure to sign your name!"

Riddle Time

Here are some colorful riddles for you to solve.

I am not small.
I am blue.
I have a blue feather.
Which hat am I?

I am not small.
I am not orange.
I have an orange feather.
Which hat am I?

I am small.
I am not orange.
I do not have an orange feather.
Which hat am I?

I am not big.
I am orange.
I do not have a blue feather.
Which hat am I?

Make up your own riddle about one of the hats.

MORE FUN. Close your eyes, point to a hat, and ask questions of others in order to identify it. Possible questions might be, "Is it big? Is it orange? Is the feather blue?" When you have guessed the hat, describe it and then open your eyes to see if you were correct.

PTERODACTYL TUNNEL

Jim stepped into the boat and grabbed the wheel. The boat rocked in the blue river. Whoosh! The boat moved forward into the tunnel. For a while Jim kept the boat going straight.

MATH FOCUS: PATTERNS. Children explore branching patterns and solve problems by using information about themselves.

Help children determine which branch they would go on at each intersection of the ride on pages 28 and 29.

Then the river divided into two branches. Jim had to decide whether to go left or to go right.

← IF YOUR NAME HAS 5 LETTERS OR LESS TURN LEFT

→ IF YOUR NAME HAS MORE THAN 5 LETTERS TURN RIGHT

Which way did Jim turn?

Which way would you turn?

MORE FUN. Make up your own pairs of directions for each intersection of the ride, for example: *If you have a sister turn left. If you do not have a sister turn right.*

21

Jim turned left and had just straightened the wheel when he heard a thunderous roar. There on his left was a gigantic stegosaur! Surprised, Jim held the wheel tightly and got past the spiky-tailed creature as fast as he could.

IF IT'S
NOON OR BEFORE NOON
TURN LEFT

IF IT'S
AFTER NOON
TURN RIGHT

Jim read the sign at the next branching of the river. He knew that it was about 2:00. Which way will Jim turn?

Which way would you turn if you were on the ride right now?

After a sharp right turn Jim looked up to see 3 apatosaurs waving their long necks at him. Feeling brave now, he waved back at them.

IF YOU ARE
10
OR OLDER
TURN LEFT

IF YOU ARE
YOUNGER THAN
10
TURN RIGHT

The river branched again, and Jim read the sign on each wall. His tenth birthday was last week. Which way will he turn?

How old were you on your last birthday? Which way would you turn?

As Jim turned left he saw what he was afraid he might see—a tyrannosaur! This time Jim didn't wave.

IF YOUR SHOES ARE WHITE TURN LEFT

IF YOUR SHOES ARE NOT WHITE TURN RIGHT

Jim read the signs and looked down at his red sneakers. Which way will he turn?

Which way would you turn if you were on the ride right now?

27

Jim steered past a pterodactyl who opened its beak and said, "Keep going out of the tunnel and into Pterodactyl Lake. Have fun steering your boat all the way to the dock."

GO INTO LAKE HERE

This is the ride Jim took. Find where Jim started to go through the tunnel. Then tell about his ride. Tell when he turned left or right and what he saw along the way. Then pretend you are going to take a ride. Decide which turns you will take at points 1, 2, 3, and 4.

ENTER HERE

DIRECTIONS FOR RIVER BRANCHES

1 If your name has 5 letters or less turn left. If your name has more than 5 letters turn right.

2 If it's noon or before noon turn left. If it's after noon turn right.

3 If you are 10 or older turn left. If you are younger than 10 turn right.

4 If your shoes are white turn left. If your shoes are not white turn right.

Showtime

Minute MYSTERY

The Terror Tour lets in only 5 people at a time.
They let groups in every 5 minutes.
They just let a group in at 1 o'clock.
What time will the next group go in?

Will the girl in the striped shirt be in the same group as the person behind her?

What time will the boy in the polka-dot sweater enter the Terror Tour?

MATH FOCUS: TIME—MINUTE AND DIVISION READINESS. By solving problems involving time, children become familiar with five-minute intervals.

Children also get experience in dividing a large group into smaller, equally-sized groups.

TERROR TOUR
Leaves every 5 minutes
5 people per tour group
Cost: 2 tickets per person

ENTER HERE

How many tickets does the ticket woman collect from each group?

What time will it be when the last person in line goes in for the Terror Tour?

If no one else gets in line, how many groups will go in altogether?

MORE FUN. Answer the questions on these pages based on the following information: The Terror Tour lets in only 6 people at a time. They let groups in every 10 minutes. They just let a group in at 1:30.

Great American Racing Derby

MATH FOCUS: PATTERNS. By recognizing, describing, and extending the pattern of winning animals on a carousel ride, children develop the ability to classify and organize information.

Have children read each list of winning animals out loud to "hear" the pattern.

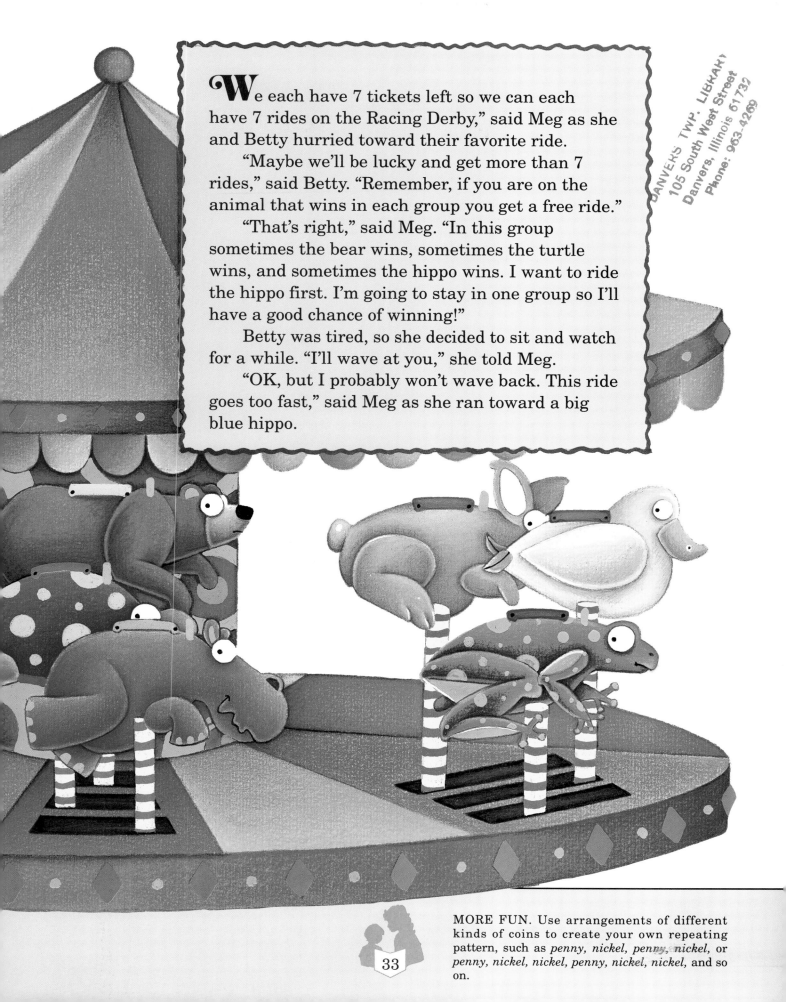

"We each have 7 tickets left so we can each have 7 rides on the Racing Derby," said Meg as she and Betty hurried toward their favorite ride.

"Maybe we'll be lucky and get more than 7 rides," said Betty. "Remember, if you are on the animal that wins in each group you get a free ride."

"That's right," said Meg. "In this group sometimes the bear wins, sometimes the turtle wins, and sometimes the hippo wins. I want to ride the hippo first. I'm going to stay in one group so I'll have a good chance of winning!"

Betty was tired, so she decided to sit and watch for a while. "I'll wave at you," she told Meg.

"OK, but I probably won't wave back. This ride goes too fast," said Meg as she ran toward a big blue hippo.

MORE FUN. Use arrangements of different kinds of coins to create your own repeating pattern, such as *penny, nickel, penny, nickel,* or *penny, nickel, nickel, penny, nickel, nickel,* and so on.

Betty watched as Meg rode around and around. After each ride Meg got off one animal, paid a ticket, and got on another animal in the same group.

This is how Meg's first 4 rides ended.

Meg's first ride

Meg's second ride

Meg's third ride

Meg's fourth ride

Which animal in Meg's group won the first race?
Which animal in her group won the second race?
Which animals won the third and fourth races?
How many of her 7 tickets does Meg have left?

Meg rode in 3 more races. She didn't win any of them. But Betty kept a record of which animal in Meg's group was the winner each time. Then she looked for a pattern of winners.

After her seventh ride a discouraged Meg came over to sit with Betty, who surprised her by jumping up and shouting, "That's it! I figured it out! I can tell from the pattern which animal will win next. Yippee!" Betty gave Meg a ticket and told her which animal to ride.

Race	Winner
1	bear
2	turtle
3	hippo
4	bear
5	turtle
6	hippo
7	bear

What pattern did Betty see? Which animal in Meg's group did Betty tell Meg to ride on? What will Meg win if the animal she rides wins the race?

35

eg ran back to the Racing Derby and jumped on the purple turtle. She couldn't wait to win the race and get a free ride.

Do you know which animal in Meg's group will win the next race?

Meg and Betty kept this record of winning animals in two other groups. See if you can figure out which animal will win next in each group. Look for a pattern.

Group 1

chicken

chicken

skunk

porpoise

chicken

chicken

skunk

porpoise

chicken

Group 2

duck

frog

duck

rabbit

duck

frog

duck

rabbit

duck

Rainbow PARK

Here's a map of Rainbow Park.

If you are at the Ferris Wheel and you want to go to Flying Saucers you could walk along Blue Bypass to Loop the Loop and then walk on Purple Parkway to Flying Saucers. How else could you get from the Ferris Wheel to Flying Saucers?

FERRIS WHEEL

GREEN ST

PINK PLACE

OVER THE FALLS

Tell three ways to go from Loop the Loop to The Twister. Which way is the longest?

BROWN BOULEVARD

What is the shortest way to go from Over the Falls to Laugh in the Dark? What's a longer way?

THE TWISTER

SUN

LAVENDER LANE

MATH FOCUS: SPATIAL SENSE—TOPOGRAPHY (MAP DIRECTIONS). By using a map to solve problems, children develop their sense of the geometric relationships of objects in a given space.

38

Children can use a piece of string to check their guesses about the longest and shortest ways to get from one ride to another.

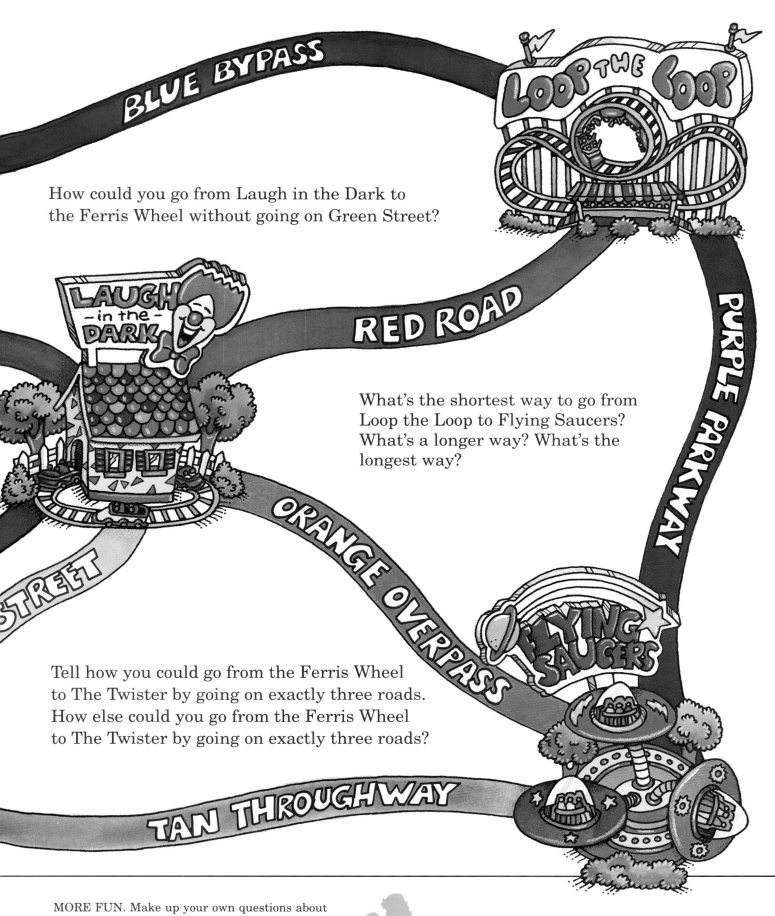

How could you go from Laugh in the Dark to the Ferris Wheel without going on Green Street?

What's the shortest way to go from Loop the Loop to Flying Saucers? What's a longer way? What's the longest way?

Tell how you could go from the Ferris Wheel to The Twister by going on exactly three roads. How else could you go from the Ferris Wheel to The Twister by going on exactly three roads?

39

MORE FUN. Make up your own questions about Rainbow Park and challenge others to solve them. Questions can include terms such as *shorter, shortest, longer, longest, exactly,* and *without.*

MATH FOCUS: SPATIAL AWARENESS, NUMBERS, NUMBER PATTERNS, ADDITION, AND SUBTRACTION. By solving a variety of visual problems, children get experience with several different areas of mathematics.

40

IT'S AMAZING

Go from START to FINISH without being stopped.

This is a 3D maze. You can follow the tunnels that go behind others.

MORE FUN. Draw your own versions of these games and challenge others to solve them, for example: Draw Number Mountain with 4 different numbers missing.

Make each side of the triangle add up to 20.
Use the numbers 1, 2, 3, 4, 5, 6, 7, 8, and 9.
Use each number only once.
The numbers 3, 4, 5, 6, and 9 are already used.

BRICK DANCE

A

B

C

D

E

F

1

2

3

Pick out the falling wall top that fits into wall 1 at the bottom.

Pick out the falling wall top that fits into wall 2.
Pick out the falling wall top that fits into wall 3.

TRIANGLE TOWER

Pick out the missing number for each blank hexagon.

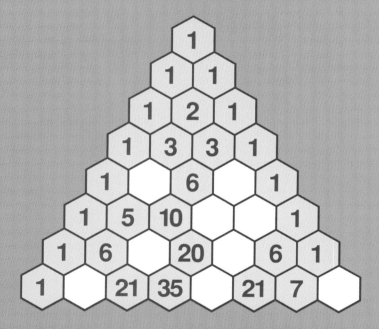

```
                    1
                 1     1
              1     2     1
           1     3     3     1
        1     _     6     _     1
     1     5    10     _     _     1
  1     6     _    20     _     6     1
1     _    21    35     _    21     7     _
```

[1] [15] [4] [10] [35] [15] [5] [7] [4]

Use each number at the bottom of the screen once.

44

Think about the row above each blank
hexagon to figure out the pattern.
Think about the row below each blank
hexagon to figure out the pattern.

LINE ATTACK

Find the puzzles that are made with one line.

Trace each puzzle with your finger.
You may not lift your finger up.
You may not trace over a line twice.

The TORCH
Minute MYSTERY

The person in charge of The Torch sits everyone in order. Starting with car 1, she puts 2 people in each car.

Billy is the first person in line. Julie is the second person in line. Which car will they sit in?

MATH FOCUS: ORDINAL NUMBERS AND DIVISION READINESS. By locating the positions of people in a line, children learn about ordinal numbers, such as first, fifth, and ninth. Children also get experience in dividing a large group of people into smaller, equally sized groups.

Sally is the fifth person in line. Which car will she sit in?
Will Sally sit with the person in front of her or the person
behind her?

Which car will the boy with the lollipop sit in?

Which car will the ninth person sit in?

Hanna is at the end of the line.
Will she get to ride on this turn?

MORE FUN. Answer the questions on these
pages based on the new information that the
person in charge of The Torch now puts 3 people
in each car.

GRAFFY PARK

It's time to go to Graffy Park.
We'll go at dawn and stay till dark.
We'll ride the rides all day and then,
We'll ride them all one time again.

So Jane and Fred and Auntie Dee,
And Tom and Mom and Anthony,
And Jack and I get in the car,
And drive awhile. It isn't far.

MATH FOCUS: STATISTICS AND LENGTH (HEIGHT)—FEET. Children use a bar graph to see the relationship between two different characteristics.

Help children refer to the bar graph on page 50 when answering the questions in the poem.

They stamp our hands and in we go,
To ride the rides both high and low.
Ready, steady, on your mark!
The gang is loose at Graffy Park!

Look at the next two Graffy pages,
To see our heights and all our ages.
How old is little Anthony?
Is he four or is he three?

MORE FUN. Make a height-and-age graph of
family members and/or friends.

Jane and Fred are twins, you see.
Are they older or younger than me?
Even though they're twin sister and brother,
Which one is taller than the other?

You must be four feet tall at least,
To ride The Roller Coaster Beast.
Which twin will wait? Which twin will go?
Look at the graph and you will know.

6 FEET							
5 FEET					TOM		
4 FEET				ME		MOM	AUNTIE DEE
3 FEET	JANE	FRED	JACK				
2 FEET	ANTHONY						
1 FOOT							

| 3 Years Old | 7 Years Old | 7 Years Old | 9 Years Old | 11 Years Old | 18 Years Old | 42 Years Old | 44 Years Old |

Who will ride on Starry Night?
Two people who are the same height.
I wonder who those two will be?
Check the graph and you will see.

THE CAGE

Tom would like to ride The Cage,
With someone who is half his age.
Who will ride The Cage with Tom?
Is it Jack, or is it Mom?

Children must be three feet tall,
To ride The Log Flume Waterfall.
Can Anthony go on the ride,
Or will he have to wait outside?

Look at the graph just once more,
To see who's age is 44.
Is the oldest person the tallest?
Is the youngest one the smallest?
Which of us are older than three?
Who is the closest in age to me?

51

Are you tall enough to ride
The Waterfall or The Beast?

Front and Center

It's almost time for the acrobatic show, and the seats are filling up fast!

You and your friend have 2 seats together. They are not in the red section. Can you find them?

Jim is sitting in the center seat of the front row of one of the sections. What section is he in? What row is he in? What number is on his seat?

Betty is in the yellow section. Her seat has a 5 on it. She is not in row B and she is not a chicken. What row is Betty in? What row is the chicken in?

MATH FOCUS: LOGICAL THINKING, SPATIAL SENSE, AND COORDINATES. Children use coordinate grids to locate seats in a theater.

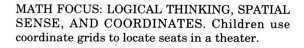

Discuss with children the arrangement of seats by color, row, and number.

How many seats are there in the red section?

How many seats are there in the whole theater?

About how many empty seats are there? Guess and then count.

Paco is looking for an aisle seat in the red section. What number is on that seat?

Who is next to the girl in the red section, row A, seat 4? What are the numbers on their seats?

Are the numbers on the empty seats in row C odd or even?

What is the row and seat number of the skunk?

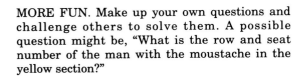
MORE FUN. Make up your own questions and challenge others to solve them. A possible question might be, "What is the row and seat number of the man with the moustache in the yellow section?"

On the Midway

Mary and her two friends are spending the day at Adventure Park. They have just finished their ride on the Chill 'N' Thrill, and now they are walking along the midway. The first booth they come to is the Mid-Weigh Balance Booth.

Mary decides to play. She has to look at all three balance scales and put the objects in order from heavy to light.

Which is the heaviest object?

Which is the lightest?

Which one weighs less than the heaviest but more than the lightest?

MID-WEIGH

MATH FOCUS: WEIGHT, LOGICAL THINKING, ADDITION, MONEY, AND COMBINATIONS. By solving a variety of visual problems, children explore several different areas of mathematics. To help children solve the problems, have them make statements about what they see, for example: On pages 54 and 55 the brick is heavier than the balloon. The cassette player is heavier than the brick. The balloon is heavier than the feather.

BALANCE BOOTH

MORE FUN. Choose a booth, make up a question about the activity pictured, and challenge others to solve it.

ANSWERS. Page 54: heaviest—cassette player, lightest—feather; Page 56: A Greeb has a rectangular body, 6 arms, and 3 hairs; Pages 58–59: 22, bear; 3+3+1, chicken; 12, dog; Page 60: Mary is in the middle; mask on left: 40¢; mask on right: 40¢.

Mary and her friends walk down the midway until they come to the Grab the Greeb booth. Mary gets 3 cards. She must figure out what all the creatures that are Greebs have in common. Then she must figure out which of her cards have Greebs on them.

Which of the creatures on Mary's cards are Greebs?

These creatures are Greebs.

These creatures are not Greebs.

22 POINTS

11~21 POINTS

1~10 POINTS

Mary and her friends walk over to the Milk Bottle Madness booth. Mary pays for two balls. She must knock down as many bottles as she can with both balls. Then she adds up the points on the bottles she has knocked down and gets a prize for that many points.

How many points would Mary get if she knocks down all the bottles? What prize would she get?

How can Mary get
a prize worth 7 points?
What prize would she get?

How many points would Mary get
if she knocks down the top bottle,
both middle bottles, and one bottom bottle?
What prize would she get?

Tell what other bottles Mary could knock down,
how many points she would have, and what prize
she would get.

59

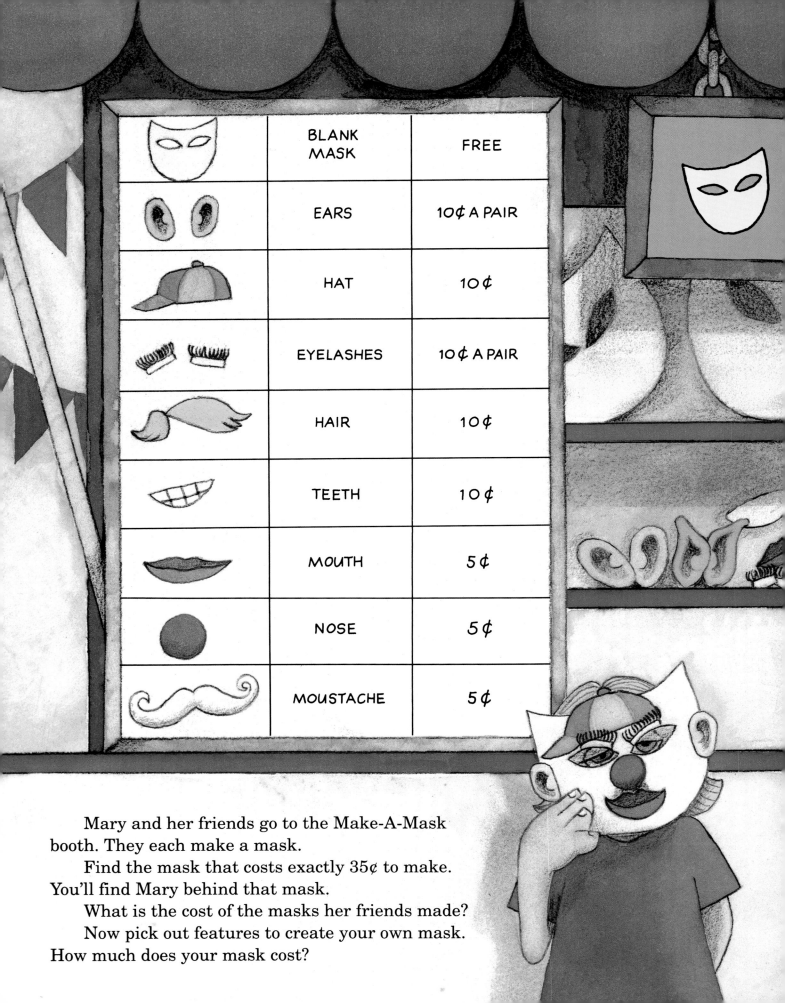

	BLANK MASK	FREE
	EARS	10¢ A PAIR
	HAT	10¢
	EYELASHES	10¢ A PAIR
	HAIR	10¢
	TEETH	10¢
	MOUTH	5¢
	NOSE	5¢
	MOUSTACHE	5¢

Mary and her friends go to the Make-A-Mask booth. They each make a mask.

Find the mask that costs exactly 35¢ to make. You'll find Mary behind that mask.

What is the cost of the masks her friends made?

Now pick out features to create your own mask. How much does your mask cost?

MAKE-A-MASK

The CALCULATOR

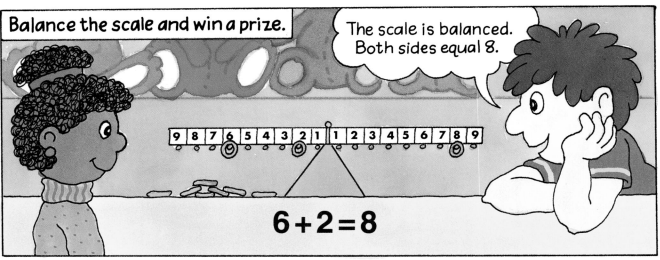

MATH FOCUS: ADDITION AND SUBTRACTION. Children use a calculator to make both sides of a number sentence equal.

For each number sentence, tell children to guess some numbers and then to use the calculator to check if their guesses were correct.

BALANCE BOOTH

Put 2 rings on the left side to balance the scale. You may put more than 1 ring on the same number.

☐ + ☐ = 8

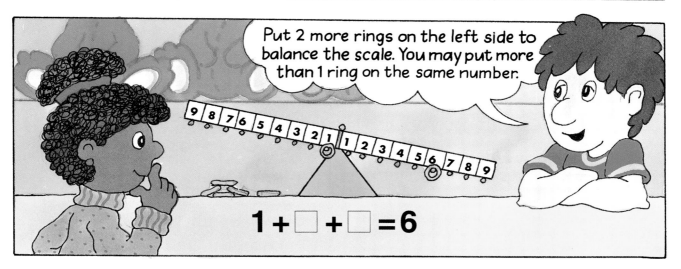

Put 2 more rings on the left side to balance the scale. You may put more than 1 ring on the same number.

1 + ☐ + ☐ = 6

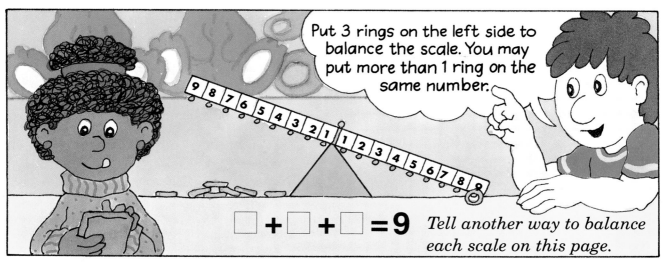

Put 3 rings on the left side to balance the scale. You may put more than 1 ring on the same number.

☐ + ☐ + ☐ = 9

Tell another way to balance each scale on this page.

MORE FUN. Use a calculator to find solutions for the following: ☐ + ☐ + ☐ = 20.

TIME-LIFE for CHILDREN®

Publisher: Robert H. Smith
Associate Publisher and Managing Editor: Neil Kagan
Assistant Managing Editor: Patricia Daniels
Editorial Directors: Jean Burke Crawford, Allan Fallow,
   Karin Kinney, Sara Mark, Elizabeth Ward
Director of Marketing: Margaret Mooney
Product Managers: Cassandra Ford,
   Shelley L. Schimkus
Director of Finance: Lisa Peterson
Financial Analyst: Tim Brown
Publishing Assistant: Marike van der Veen
Administrative Assistant: Barbara A. Jones
Production Manager: Marlene Zack
Production: Celia Beattie
Supervisor of Quality Control: James King
Assistant Supervisor of Quality Control:
   Miriam Newton

Produced by Kirchoff/Wohlberg, Inc.
866 United Nations Plaza
New York, New York 10017

Series Director: Mary Jane Martin
Creative Director: Morris A. Kirchoff
Mathematics Director: Jo Dennis
Designer: Jessica A. Kirchoff
Assistant Designers: Daniel Moreton, Judith Schwartz
Contributing Writer: Anne M. Miranda
Managing Editor: Nancy Pernick
Editors: Susan M. Darwin, David McCoy

Cover Illustration: Tom Leonard

First printing. Printed in U.S.A.
Published simultaneously in Canada.

Time Life Inc. is a wholly owned subsidiary of THE TIME INC. BOOK COMPANY

Time-Life is a trademark of Time Warner Inc. U.S.A.

For subscription information, call 1-800-621-7026.

CONSULTANTS

Mary Jane Martin spent 17 years working in elementary school classrooms as a teacher and reading consultant; for seven of those years she was a first-grade teacher. The second half of her career has been devoted to publishing. During this time she has helped create and produce a wide variety of innovative elementary programs, including two mathematics textbook series.

Jo Dennis has worked as a teacher and math consultant in England, Australia, and the United States for more than 20 years. Most recently, she has helped develop and write several mathematics textbooks for kindergarten, first grade, and second grade.

Dr. Carole Greenes, an internationally recognized math educator, is a professor of mathematics education at Boston University, where she is also associate dean of research, development, and advanced academic programs for the School of Education. She is noted for her interest in mathematical problem solving, the use of computers in education, and techniques for helping special-needs students learn math. She is a frequent speaker at national and international meetings and has written and collaborated on many books, including two major mathematics textbook series.

Illustration Credits: Bob Barner, pp. 48–51; Liz Callen, pp. 30–31, 52–53; Tom Leonard, pp. 20–29, 40–45; Don Madden, pp. 12–19; Daniel Moreton, front end papers, pp. 32–37; Carol Nicklaus, pp. 46–47, 62–63; Andy San Diego, pp. 10–11, 54–61; Lou Vaccaro, pp. 38–39; Joe Veno, pp. 6–9; Troy Viss, pp. 41–42, back end papers.

Photography Credits: Pages 40–45, John Lei/OPC.

Library of Congress Cataloging-in-Publication Data
Pterodactyl tunnel : amusement park math.
        p.   cm. — (I love math)
    ISBN 0-8094-9990-8
    1. Mathematics—Juvenile literature. 2. Amusement parks—Juvenile literature. [1. Mathematics. 2. Amusement parks.] I. Time-Life for Children (Firm) II. Series.
QA40.5.P84   1993
510—dc20   93-31567
                                                        CIP
                                                        AC

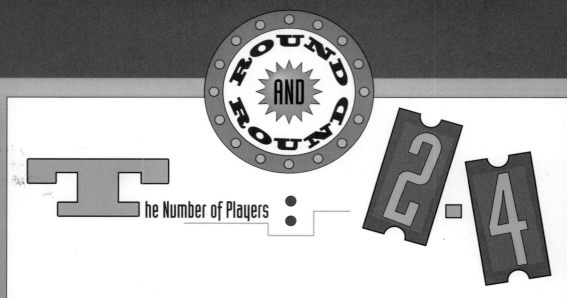

The Object of the Game: To collect as many counters as possible while going all the way around the Ferris Wheel once.

The Playing Pieces: Counters, such as cereal pieces or pennies; a single die; a different kind of playing piece for each player.

The Play: Each player starts with 10 counters. An additional 20 counters are put in a pile at the center of the Ferris Wheel. Players throw the die and the one with the lesser number goes first. Players take turns throwing the die and moving that number of cars clockwise. If the number thrown is odd, the player takes that number of counters from the pile. If the number is even, the player puts back that number of counters. If a player lands on a car already occupied, the player who was already there goes back to Start.

The play continues until all players have gone around the Ferris Wheel once. The players must arrive at car 1 on an exact count. Until the exact count is thrown, a player continues to take and give back counters. After all players have returned to car 1, they add up their counters.

The Winner: The player who collects the most counters is the winner.

Math Concepts: Odd and even numbers.
Number comparisons.